I

HISTORICAL

The object of this paper is to present the subject of nutrition in its broad general aspects and to suggest the possibility of the practical application of some of the facts which years of labor through many generations of workers have brought to light.

It seems as though mankind had a right to a knowledge of the value of the foods which a bountiful Nature has provided for his use. Even among educated persons one may hear the grossest errors of judgment regarding the nutritive value of a hen's egg and few of those who eat in restaurants realize that the greater quota of nourishment which is brought to them lies not in the specific dish served but in the bread and butter which ostensibly is presented as a gift.

From the earliest times it was evident that although an adult partook of a great deal of food, he did not gain in weight. Hippocrates believed this to be due to a constant loss of insensible perspiration and to the elimination of heat, which he conceived to be a fine form of matter. Galen, six hundred years later than Hippocrates, was no further advanced in his conception of nutrition. For thirteen hundred years after Galen intellectual progress lay dormant under the spell of the Dark Ages.

One of the first inquirers of the Renaissance, the brilliant Paracelsus, explained the phenomenon of nutrition as being under the supervision of an archeus, a spirit which dwelt in the stomach and separated the food into the good and the bad, the good being used by the organs of the body, and the bad eliminated.

A true conception of the nutritive process could only be formulated when a knowledge of the existence of the various gases was revealed. It was Lavoisier who first showed that when an organic substance burned, the products of combustion were equal to the sum of the original substance and oxygen. Oxygen had but recently been discovered by Priestley. Lavoisier burned plants and found that carbon dioxide and water resulted. He, therefore, concluded that they contained carbon and hydrogen. Animals contained nitrogen in addition. This was the first analysis of organic material.

Lavoisier went further and found that an animal or a man, like a burning piece of wood, absorbed oxygen and eliminated carbon dioxide. He discovered that the process of heat production in man was one due to oxidation, that the prevailing idea that particles of air entered the salt and sulphur containing blood and there caused fermentation was untrue. Lavoisier measured the heat given off by a guinea pig by noting the quantity of ice melted by the animal when placed in a hollow block of ice, and he measured the gases given off by the animal in order to determine whether the heat produced could be accounted for by the oxidation going on.

He, furthermore, determined that oxidation in man was increased by giving him food, by causing him to do mechanical work or by subjecting him to the influence of cold. Reflecting upon these facts during the troublous times of the French Revolution, Lavoisier wrote, "Does it not seem a great injustice of Nature that the poor laborer uses more of his body substance, while superfluity, which is unnecessary for the rich, should be his portion?"

To the darkness of the history of the time belongs the fact that Lavoisier, begging, according to Carlyle, for two weeks more of life in order to complete his experiments, was guillotined, thereby becoming the greatest sacrifice of the insensate fury of his age. (For this earlier literature see Carl Voit: Ueber die Theorien der Ernährung der tierischen Organismus, München, 1868.)

The progress of science is a history of great discoveries of fact which become established, and of destruction of theories which are temporary mental conclusions shown later to be untenable. Nor can a master mind like that of Lavoisier escape the application of this universal law. He showed that animal heat was due to a process of oxidation but he believed that the heat produced was caused by the union of oxygen with carbon and with hydro-

gen in the lungs. It was not till sixty years after
his death that it was fully realized that the heat production was due to the oxidation of protein, fat and
carbohydrate within the different organs of the
body.

Carl Voit, to whom more than anyone else the
world owes its fundamental knowledge of nutrition,
was accustomed to say in his lectures, "Continual
decompositions of matter are always going on in the
living cells, and the energy liberated in these decompositions is the power upon which the motions of
life depend. Phenomena of life are phenomena of
motion." In truly poetical language Rubner, the
most eminent of Voit's pupils, has written, "Mute
and still, by night and by day, labor goes on in the
workshops of life. Here an animal grows, there a
plant, and the wonder of it all is not the less in the
smallest being than in the largest."

The workshops of life require fuel to maintain
them, and a necessary function of nutrition is to
furnish fuel to the organism that the motions of life
continue. Furthermore, the workshops of life are
in a constant state of partial breaking down and
materials must be furnished to repair the worn-out
parts. In the fuel factor and the repair factor lie
the essence of the science of nutrition.

These two factors operate to bring about death
from starvation, either the body's own store of fuel
becomes exhausted or a part of the machinery

necessary for life wears out. As regards the course of death from starvation, there exists the written record of the explorer Hubbard. The following words are believed to have been penned a few hours before his death in Labrador. "I am not suffering. The acute pangs of hunger have given way to indifference. I'm sleepy. I think death from starvation not so bad. But let no one suppose I expect it. I am prepared—that is all." Hubbard's biographer quotes the following as showing the spirit of the lost explorer, as it indeed represents the spirit of all investigators:

" Something hidden. Go and find it. Go and look behind
the Ranges,
Something lost behind the Ranges. Lost and waiting
for you. Go."

II

THE CONSTANT NEED OF FUEL

The light and heat of the sun playing on the green leaf of the plant cause carbon dioxide and water to unite to form sugar. Heat is absorbed in the process and oxygen is given off to the atmosphere. If one gram of sugar be placed in a very strong, closed steel receptacle into which oxygen, under a pressure of 450 pounds to the square inch, is conducted, and then if the sugar be kindled by an electric spark it will be completely burned to carbon dioxide and water and exactly the same quantity of heat will be liberated as was obtained from the sun in the original manufacture of the substance. If the steel receptacle be placed in a liter of water in which a thermometer has been put it may be noticed that the temperature of the water rises nearly 3.75°. After making certain corrections, it may be proved that 1 gram of glucose, when oxidized, yields heat sufficient to raise one liter of water 3.755° C. Since the measure for heat is a calorie or that quantity of heat required to raise 1 liter of water 1° C., it follows that 1 gram of glucose yields 3.755 calories of heat. This apparatus measures the heat of combustion of organic substances like sugar, starch, meat, fat, etc., and is called a bomb calorimeter (a measurer of calories). The calorie or heat unit is as much an exact value for measurements of heat as are a quart or a pound for measurements of volume or weight.

When protein is burned in the bomb, the nitrogen of it is converted into nitric acid, but when protein is burned in the body, its nitrogen is not oxidized but is eliminated in the form of urea, so the heat produced from protein in the body is always less than that measured by the bomb. Sugar, however, yields the same products in the bomb and in the body and, therefore, the amount of heat produced is identically the same, no matter where the oxidation takes place. The same is true of fat.

One gram of the ordinary food stuffs when oxidized in the body yields the following number of calories:

	Calories
Glucose	3.755
Cane sugar	4.0
Starch	4.1
Fat	9.3
Protein	4.1

It has been said by some that they never will be converted to the belief that a knowledge of calories in nutrition is valuable. These persons must be reasoned with and persuaded to listen and they cannot then but be convinced.

The law of the conservation of energy holds that power cannot arise from nothing. Power must be derived from some store of energy, from energy which is potential. The store of power in the food stuffs is liberated when they are oxidized in the body. This power becomes the source of the motions of life, and in the resting organism is finally liberated as heat. If this be true, then if one can measure the quantity of protein, fat and carbohydrate (sugar) oxidized in an animal in twenty-four hours one can calculate the quantity of heat which will arise from this process. If, at the same time, the animal can be placed in a calorimeter, which measures the heat actually given off during the period, the two computations should exactly agree.

To Rubner belongs the glory of being the first to have demonstrated this truth.

Calorimeters have since been constructed to measure the heat production in man. The labors of Atwater, Rosa and Benedict have confirmed the application of the law of the conservation of energy to man. The oxygen intake and carbonic acid outgo give a measure of the oxidation of the food stuffs, and the heat given off by the body is found to be equal to the quantity of heat which would have arisen from the oxidation of just that quantity of protein, fat and carbohydrate estimated to have been destroyed.

In a calorimeter built for the Russell Sage Institute of Pathology in Bellevue Hospital, it has been found that if a definite amount of alcohol be burned in a lamp within the apparatus, the heat measured during a four-hour period is exactly that amount which the theory would call for.

Theory	Found
212.57	211.88

Doctors Du Bois and Warren Coleman have discovered that a typhoid patient, during a period of five hours of rest in this calorimeter, produced the same number of calories as were calculated he should produce from the materials which were oxidized in his body.

Theory	Found
422.59	419.78

Contemplation of such a result as this drives home the fact that if this typhoid patient is to be kept from losing his own body muscle and fat, he must be given the equivalent of 422 calories in food substances during a five-hour period.

If one measures the hourly heat production of a normal resting man, one must be convinced of its constancy. The human furnace requires a certain quantity of fuel to support the activities of life. Measurement of the total heat production, therefore, becomes a measurement of the intensity of the life processes.

The quantity of heat produced by mammalia of the same size is fixed and definite and can be closely predicted in advance. It is not dependent on the weight of the animal nor upon the relative size of the individual cells. Thus, the size of the cells which make up the substance matter of a mouse is not very different from the size of the cells of the horse, yet a mouse produces 452 calories per kilogram of body weight in 24 hours and the horse 14.5 calories. The mouse requires thirty times more food per unit of body weight than the horse. However, Rubner has shown that all well-nourished mammals produce the same number of calories per square meter of surface.

A normal man, well nourished, who is resting quietly in his bed in the morning, having been without food for fifteen hours, will manifest a minimum level of heat production. This level may be called the *basal heat production*. The following table has been prepared to show the constancy of energy production under these circumstances:

Individual	Weight in kilos	Calories per hour per kilo	Calories per hour per square meter surface
B.	83	1.01	35.7
G. L.	78	1.03	36.5
D. B.	74	1.01	35.3
R.	74	0.95	32.5
C.	68	0.96	31.7
J. R.	66	1.00	32.7
H.	62	1.15	37.0
G.	56	1.07	34.0
T. C.	49	1.13	37.7

In the light of this exposition no educated man can say that he "does not believe in calories," when the energy in the food stuffs constitutes the basis of his being, and calories eliminated from his body are a measure of the sum total of his physical activities.

Food is the fuel of the human furnace, and must be furnished to that furnace in accordance with its needs.

The basal heat production of an average man weighing 156 pounds (70 kg.) will be 70 calories per hour or 1,680 calories in twenty-four hours. If food be taken extra heat is produced in the body. This extra amount does not exceed 10 per cent of the basal heat production or 7 calories per hour and 168 per day, so that the maintenance requirement of this man, resting quietly in bed, would be 1,848 calories in the daily diet.

Beyond this, the amount of fuel needed depends upon the quantity of mechanical work done. It becomes purely a matter of supplying fuel for the machinery.

It has been shown by Atwater and by Benedict that if a person sits in absolute quiet in a chair the heat production is 8 per cent greater than when he is lying on a bed. If, however, those ordinary movements are made which are associated with daily life when sitting in a chair, the heat production may rise 29 per cent, or from a basal level of 70 to a level of 90 calories per hour, an increase of 20 calories.

Since the influence of food is to increase the metabolism 7 calories per hour during twenty-four hours, and the influence of a sedentary life adds 20 more calories per hour during the 16 hours when a man is up in his chair, the total energy requirement would be:

	Calories
Night $(70+7) \times 8 =$	616
Day $(70+7+20) \times 16 =$	1,552
	2,168

A hospital patient must be liberally fed when he receives this amount during convalescence. Additional fuel in the food may be considered as a charitable contribution, a welfare fund for future use.

No normal man leading a life involving sedentary occupation, should live without exercise. Only this will keep his body in proper condition. One may attempt to calculate the additional energy requirement needed for this purpose. To walk one hour on a level road at the rate of 2.7 miles requires energy to the amount of 160 additional calories. Therefore, if the man of sedentary occupation walks two hours daily to and from his business, 320 calories must be added to the 2,168 required to support him without exercise, a total of 2,488. This figure is not far from Rubner's average allowance of 2,445 calories for such men as writers, draughts-men, tailors, physicians, etc.

It follows, therefore, that about 2,500 calories are required in the daily food of a man whose occupation is of sedentary character. As a matter of fact, statistics show that the inhabitants of cities take this amount of fuel daily. The latest statistical proof that the food supply of a great city is regulated by the needs of its inhabitants may be found in the report of Gautier which shows that in Paris an average of 2,500 calories of energy are daily supplied to each inhabitant.

If the exercise taken be vigorous and include hill climbing, the quantity of energy needed will be greater than when a level road is traversed. To climb on a path at the rate of 2.7 miles an hour so that the summit of a hill 1,650 feet high is attained during the hour, requires 407 extra calories.

An expert bicycle rider at hard work has indicated an increase in oxidation corresponding to 529 calories per hour (Atwater and Benedict).

The fuel requirement, therefore, depends upon the quantity of work accomplished.

The Kaiser Wilhelm Institut has recently granted funds to Rubner in Berlin in order to establish a special laboratory in which to determine the specific fuel needs of individuals engaged in various occupations and trades.

Computations of the diets of farmers show the

following interesting similarity in the fuel values of their food:

	Calories
Farmers in Connecticut	3,410
Farmers in Vermont	3,635
Farmers in New York	3,785
Farmers in Mexico	3,435
Farmers in Italy	3,565
Farmers in Finland	3,474
Average	3,551

These figures, representing the food-fuel contained in the dietaries of individuals in widely differing communities but engaged in the same occupation, show a plus or minus variation of only 6 per cent from the mean average.

From the present available data one may estimate the daily fuel requirement of well-nourished adults after the following fashion:

Occupation	Calories
In bed 24 hours	1,680
In bed 8 hours, work involving sitting in a chair 16 hours	2,170
Bed 8 hours, in a chair 14 hours, moderate exercise 2 hours	2,500
Farmers	3,500
Rider in a six-day bicycle race	10,000

It is apparent that the great numbers of men employed as clerks or those employed in watching machinery require about 2,500 calories in their daily food.

A boy of twelve requires about 1,500 calories daily. A baby when first born requires 100 calories per kilogram of body weight per day and later about 70. Many cases of reported chronic malnutrition of infants are in reality due to persistent undernutrition carried out in ignorance of the proper amount of food required by the child.

In fever, the production of heat may be 50 per cent above the normal. In cases of hyperthyroidism (Graves disease) even greater increases have been observed, whereas in hypothyroidism (myxœdema) the heat production falls below the normal. It follows, therefore, that increased nourishment is indicated in fever and in Graves disease whenever this is possible.

The great practical importance of food fuel in sufficient quantity for the human machine in health and disease warrants its consideration in greater measure than has heretofore been given it.

III

THE CONSTANT NEED OF PROTEIN

If a man take only fat, sugar and starch in his diet, he will be unable to maintain his life during a long period. Voit tells how an English physician nourished himself with sugar alone for a month, became extremely weak and shortly thereafter died, a victim of his scientific curiosity. A diet may be deficient in calcium salts and, therefore, the body may suffer from calcium hunger. Or a diet may be poor in iron, as a milk diet is, and the person nourished on such a diet may become anemic from want of sufficient iron to form new red blood cells.

However, those who live on the usual mixed diet rarely suffer from salt hunger. Ample quantities of salts are found in milk, and iron is present in the yolk of eggs, in meat and in green vegetables, especially in spinach. Salts, therefore, scarcely enter into the food as an economic question. Common table salt is purchased but its use is largely that of a flavor. When potatoes are taken common salt becomes a physiological necessity, but its ordinary use is in excess of the amount actually required.

There is, however, one important material to be treasured and protected and that is body protein. There are different kinds of proteins, such as those of milk, meat, gelatin, fish, vegetables. In the process of digestion all of these different kinds of proteins are broken up into a great number of nitrogen-containing acids. These have been compared in number to the letters of the alphabet. When they are arranged together they can make many different proteins just as there are many different words in the dictionary.

Suppose the word albumin were broken up by digestion into the letters a, b, i, l, m, n, u, then if these letters were absorbed they could be reconstructed into albumin again. Assume the same for the word globulin. Now if both albumin and globulin were to be formed from a common word, one would have to ingest a hypothetical substance called amglobulin, convertible into globulin if the letters a, m are abandoned to their fate or into albumin on similarly exorcising the letters l, g. Carrying the analogy still further it is evident that if the letter b were not in the word amglobulin, neither albumin nor globulin could possibly be produced.

This gives a key to the physiological value of different proteins. Casein, the principal protein of milk, contains practically all the various elemental forms of those structural materials which enter into the different proteins used to make up the framework of the machinery in the living cells. .

The most important letters of the protein alphabet are the following: glycocoll, alanin, valin, leucin, prolin, phenylalanin, aspartic acid, glutamic acid, serin, tyrosin, cystin, lysin, histidin, arginin, ammonia and tryptophan.

The infant has the power of transforming 40 per cent of the protein in its food into new structural machinery, the architecture of which depends upon a regrouping of individual units formerly in the protein of milk. Thus new proteins are built whose internal arrangement is dependent upon local conditions in the various organs of the child.

Now it is apparent that proteins are especially valuable if they contain an array of units which, when reunited, form body proteins. Such food proteins are those of milk, meat, eggs and fish. It is also apparent that proteins like gelatin and zein, in which one or more of the necessary units are lacking, can never be reconstructed into new body protein. Finally, it must be clear that when the units are in a very different ratio to one another from that in which they exist in body protein, they must be of inferior nutritive value, since large quantities must be broken up in order to yield that quantity of certain units necessary for the construction of animal protein. Such inferior proteins occur among the plants. Plant proteins are eaten by the ox and are reconstructed into beef proteins, with the oxidative elimination of the excess of chemical units which are unnecessary for the structure of the animal cell. In this way beef protein attains a higher biological value for the nutrition of man than is possessed by vegetable proteins.

The body of an average man weighing 156 pounds contains about 30 pounds of protein or 20 per cent of the live weight. If the man starves he will lose 5 parts per thousand of his protein store daily. If he be given fat and carbohydrate in large quantity, the daily loss of body protein may be reduced to 2.5 parts per thousand. This loss of body protein represents the irreducible minimum of wear and tear on the constituent parts of the machinery of the cells. Murlin has shown that this minimal destruction cannot be prevented by giving gelatin with fat and carbohydrates. Gelatin contains many of the structural units of meat protein but in very different relative amounts and it contains no tyrosin, cystin or tryptophan. It, therefore, has not the chemical units necessary to repair the worn-out parts of the cell machinery. Murlin found, however, that if this quantity of protein which constitutes the *irreducible minimum* of wear and tear on the cells was added in the form of beef heart to the gelatin diet, the waste of body protein stopped at once. He found that the wheat proteins of cracker meal were far less efficient in protecting the body from protein loss than were the proteins contained in beef heart.

Thomas, in Rubner's laboratory, took starch and sugar in large quantity in his diet and determined the minimal loss of protein under these circumstances. He then took meat equal in amount to this minimal quantity destroyed and found that if the food was divided into six portions and taken four hours apart, there was no loss of body protein.

These experiments show how necessary it is that the body have a constant replenishment of its protein store. The experiments of Thomas were carried still further and showed the relative biological value of the proteins of different origin. The following minimal amounts were required to protect body protein from loss:

Meat protein	30 grams
Milk protein	31 grams
Rice protein	34 grams
Potato protein	38 grams
Bean protein	54 grams
Bread protein	76 grams
Indian corn protein	102 grams

There can be no doubt whatever as regards the superior value of meat, fish, egg and milk proteins over those of bread, beans and Indian corn. The proteins of rice and potato, however, hold an intermediate position.

Such facts as these should make it possible to classify proteins into groups according to their physiological value. Milk is sold in New York as of three grades, A, B and C. In like manner the proteins of the food stuffs could be labelled A, B and C according to their physiological value, and to group D might belong gelatin and some other proteins which cannot replace the body protein that is continually wearing away.

The question will at once be asked, Why do the vegetable proteins behave differently from meat and milk proteins? The answer is given by the work of the American investigators, Osborne and Mendel. Osborne, for years, has prepared vegetable proteins in their purest form. He has found, for instance, that the principal proteins in wheat are two in number, existing in almost equal amounts, called wheat glutenin and wheat gliadin. Glutenin, on analysis, yields about the same integral chemical substances as casein. Gliadin, the alcohol-soluble wheat protein, does not contain the unit known as lysin; and it contains 37 per cent of glutamic acid, of which substance milk and muscle protein contain only about 10 per cent. Osborne argued from these facts that wheat gliadin would not be as valuable in nutrition as milk protein. The question was, how to compare the physiological value of the different proteins. The method adopted was to feed young albino rats, just weaned, with a diet containing milk sugar, fat, milk salts and the protein to be tested, and then to observe the curve of growth of the rats and see if it were normal. When casein was the protein added to the diet, the rats showed a usual rate of growth and lived a year and more without the slightest abnormality. It was found that the growth took place freely when 15 per cent of the calories in the diet were in the form of protein. With white mice, on the contrary, nearly 25 per cent of the calories in the diet had to be in protein, if the normal development of these smaller and more rapidly growing organisms was to be provided for. These figures may be contrasted with 7 per cent of calories in protein in human milk, which adequately provides for the growth of the infant. The diet of rats required 3 per cent of milk salts whereas mice required 6.8 per cent for proper growth. These, therefore, are factors of considerable importance.

Not only casein but also lactalbumin of milk, ovalbumin of egg, edestin of the hemp seed, and glutenin of wheat, were able, each in itself, to be an all-sufficient source of protein supply when furnished with the diet of fat carbohydrate and milk salts above described.

Wheat glutenin, therefore, is an adequate protein providing for maintenance and growth. However, when gliadin from wheat was the protein given to rats, growth was almost inhibited. The weight was maintained but the animals were dwarfed. The capacity for normal growth was present, for addition to the diet of an adequate protein like casein caused normal growth at any time, even though the ch:
gr<

of normal nutrition. Furthermore, addition of the missing protein unit lysin to the gliadin diet caused the rats to grow. This analysis shows that gliadin, which represents nearly half the protein of wheat, is of inferior food quality not only on account of its large content of glutamic acid but also because it lacks a necessary letter of the protein alphabet, lysin. It interprets the results of Thomas, which show that it requires more protein in bread to protect the body from protein loss than it does when milk or meat are ingested. Other vegetable proteins such as hordein from barley and gliadin from rye behaved like wheat gliadin.

Indian corn contains proteins, among which are glutenin and zein, the latter constituting a little over half the protein in the corn. When corn glutenin was given as the protein in the rat's dietary, normal growth was recorded. However, when the alcohol soluble zein was given, the rats lost in weight and died unless a change in diet afforded relief. Zein is a protein which is like gelatin in that it contains no tryptophan. Such a protein can never form new tissue, nor protect body tissue from its normal wear and tear, so the rat must perforce succumb. The remarkable verification of this theory lies in the fact that the addition of the missing tryptophan to the zein diet of rats or mice, either greatly prolongs their lives or may even cause them to maintain their weight (Gowland Hopkins, Ruth Wheeler,

Osborne and Mendel). These experiments show why corn proteins are not the physiological equivalent of meat protein in nutrition.

Phaseolin is a protein found in the kidney bean and behaves like zein in nutrition, failing completely to maintain the body weight of the rats which have received it. It may, therefore, be considered a protein lacking in some chemical unit.

The New Haven investigators point out how a single protein like casein can cause normal growth involving the production of all the various body proteins, such as hemaglobin, other blood proteins, elastin, collagen, the keratin of skin and hair and so forth, and even the complex nucleo-proteins also.

Casein does not contain glycocoll but glycocoll may be manufactured from it within the animal organism. The body cannot, however, make tryptophan from gelatin or from zein, hence these can never form body protein or be used in its repair.

The proteins of rice and potatoes are scarcely known because Osborne has not yet analyzed them.

Enough has been said to give the reason why meat proteins are of greater physiological value in the diet of man than are vegetable proteins. It is evident, however, that if enough bread or enough corn be taken, a sufficient quantity of their content of the more valuable proteins will be obtained. The ox merely takes these miscellaneous and variably constituted proteins and rearranges their constituent units into his own flesh, which is in composition approximately like our own. In like manner the cow forms casein, a complete food protein, available both for maintenance and growth.

Protein is usually taken in excess of that bare requirement which is measured by the quantity necessary to repair the tissue. This excess is oxidized and used as fuel just as are fat and carbohydrates. Protein has one property out of all proportion to that possessed by the other food stuffs: it very largely increases the production of heat in the body. Individuals maintained on a low protein diet may suffer intensely from the cold. A good piece of beefsteak or roast beef will put the heat production on a higher level, and a person going out of doors on a cold day after a meal high in protein does not feel the cold. For the same reason, on a hot summer day, meat will be avoided.

One may cause the heat production of a dog to double by giving a large quantity of meat. This action is due to the fact that many of the chemical units produced in the digestion of meat act to stimulate the heat production in the animal. Some of these individual chemical units of which protein is composed have been fed to a dog and have been found to increase the heat production in the same way as does ingested meat. Glycocoll is the simplest of these units. It causes a great rise in heat production. In diabetes ingested glycocoll is completely

converted into sugar without undergoing oxidation.
When thus given in diabetes glycocoll still causes an
increased heat production although it is not oxidized.
Hence it acts as a chemical stimulus and not in
virtue of its energy content.

This heat-increasing property of protein has been
called its specific dynamic action (Rubner). This
action may be effected by all kinds of proteins, by
those of meat, fish, eggs, milk, and such incomplete
proteins as gelatin. To obtain the warming effect it
is not necessary to purchase beef, a relatively costly
article of diet.

The quantity of protein desirable in the dietary
had best be given consideration after discussion of
the dietary habits of mankind. It is sufficient for
the present to realize that protein is necessary to
maintain tissue in repair, to promote growth and
that when taken above the requirements for these
purposes it stimulates the organism to a higher level
of heat production.

IV

HABITS OF DIET

There is a wide diversity of dietary habits in different parts of the world. The inhabitants of the tropics live from the fruits which grow readily without especial cultivation. There being no reason to work hard for food and lodging, there is no incentive to struggle and no advance in civilization. The struggle for food and lodging in the zones of moderate temperature has strengthened the body and mind of man. The food of the temperate zones is usually a diet containing a mixture of animal and vegetable substances.

A race of human beings which is practically carnivorous in its habits is the Eskimo. Recent studies (August and Marie Krogh: A Study of the Diet and Metabolism of Eskimos, Copenhagen, 1913) show that the Eskimo hunter may rise early in the morning, drink a cup of water, of soup or of coffee. He then goes out without food or he may take a small bit of dried or frozen meat. Returning at three or four o'clock in the afternoon he fills himself with meat to his utmost capacity as soon as it can be cooked. He then sleeps for two hours, after which he enjoys himself socially and before retiring for the night takes a second smaller meal, usually of fish.

The principal articles of diet are seal meat and the meat of reindeer, walruses and whales. The skin of young whales is esteemed a delicacy and this has been found by Bertelsen to contain a very large quantity of animal starch (glycogen). The only native vegetables are whortleberries, young shoots of angelica, and seaweed which is eaten with mussels.

When game is plentiful large quantities of meat and fat are eaten. Rink reports that young robust people may ingest as much as four kilograms (9 pounds) of meat daily during the time when seals are plentiful. In this way the Eskimos store ample fat in their bodies. Although this people are the greatest meat eaters in the world there is practically no gout among them. Also there is no quarreling. If two persons do not like each other they simply move away from one another, but a blow is never struck. They have great capacity for physical endurance and for resistance to external cold. On sledge journeys in the north during the winter when the temperature is —30° C., they may take a full meal of frozen meat and blubber at night. The first result is a feeling of extreme cold and shivering, but after half an hour the stimulating effect of the large absorption of protein (meat) upon the heat production makes itself felt and the Eskimo may then sleep in the open with no other extra protection than his sled which is put up as a shield from the wind. The same kind of food is also taken by the dogs accompanying the party.

If one turns from the carnivorous Eskimo to the population of India, China, Japan and the Philippine Islands, one finds that the great staple of diet is rice. Rice is a material which is almost tasteless in itself but which can be prepared in many ways and may be eaten with many diverse flavors.

The poorest families of Southern India live essentially on rice to which is added a small quantity of fish. Rice can be raised with little effort but poverty excludes meat from the diet. McCay (The Protein Element in Nutrition) states that it may be accepted that these rice-eating inhabitants of Bengal, whose diet is poor in protein, are incapable of performing a really hard day's work, the explanation being an incomplete development of muscular tissue. It seems that the tropical heat of India yields nourishment as freely as it may be obtained in Greenland by the Eskimo, yet physical power is lacking in the inhabitants of the country.

Among the hill tribes of India and wherever else animal protein is taken with rice, a higher physical development is found.

The poorest Philippinos also live mainly on rice but they add sufficient fish to furnish an adequate diet. Aron (*Philippine Journal of Science*, 1909) estimates that the daily cost of this diet per person in the town of Taytay amounts to 12½ centavos or 6¼ cents.

The most celebrated European standard diet is that of Voit designed for a laboring man working hard during a period of nine or ten hours. This diet contains 118 grams of protein, 56 grams of fat and 500 grams of carbohydrates. Of the 118 grams of protein, 46 are furnished in 230 grams (one-half pound) of butcher meat. The quantity of meat in the ration was determined from the finding that each inhabitant of Munich consumed daily an average of 205 grams of meat and 25 grams of fowl and fish. The other factors were determined by statistics and they were verified by examining the diets of various laboring men. The quantity of fat was put low in the diet on account of its cost.

Volumes have been written concerning the necessity of the quantity of protein in Voit's dietary. Voit, himself, showed that a man could exist with half that quantity but in the violent debate that centered around this question Voit took no personal part.

When comparison is made of the diet of the Eskimo, the Ber   Europe, the follo nutritive elements

	Weight in kgm.	Protein grams	Total calories	Calories in %		
				Protein	Fat	Carbohydrates
Eskimo	65	282	2,604*	44	48	8
Bengali	50	52	2,390	9	10	81
European	70	118	3,055	16	17	67

The figures regarding the Eskimo are taken from Krogh, who believes that they are not 10 per cent from the actual values. The carbohydrate reported is largely that of the glycogen content of the flesh eaten. The Eskimo takes five times the amount of protein eaten by a Bengali and two and a half times the amount eaten by a European. Yet he lives without acquiring uric acid diseases (Krogh). The specific dynamic action or heat-stimulating property of protein, as well as the Eskimo's coat of subcutaneous fat enables him to bear extremes of cold. The acidosis which arises when large quantities of fat are ingested without carbohydrates, is in all probability counteracted by a large production of sugar from the fragments of broken down protein. The production of sugar from glycocoll has been remarked upon and serves as a typical illustration of this process. Health and strength, therefore, are not wanting when in the Arctic regions a carnivorous diet is the mainstay of life.

---

*Statistical average. The quantity of seaweed and mussels eaten could not be controlled.

As regards the Bengali, it seems that his ration of rice and fish does not serve to maintain him in good condition, whereas his neighbors in the hills who partly live upon the flocks they raise are in much better physical condition.

Voit's ration contains about four times the minimal quantity of protein necessary for the maintenance of life. Voit's dietary has been condemned as financially extravagant and even physiologically harmful. Evidently many millions of dollars could be saved in the army and navy were the protein of the ration cut in two. Rubner recently appeared twice before the German authorities to protest against such reduction. He believes that there should always be an excess of protein constructive material, so that if after physical exhaustion there is depletion of the glycogen reserves, under which circumstances the wear and tear on the cell protein is increased, there may be building units in reserve to quickly restore the tissue destroyed.

For a laboring man to take the minimal quantity of protein in the diet is, therefore, not desirable. Meltzer has truly pointed out that eating protein in quantities above the minimal requirement is one of the many "factors of safety" in human life.

Moreover, it would be a difficult proposition to arrange a low protein dietary for such laboring men as farmers, for example. Rubner points out that the staple diet of the robust and healthy Bavarian peasant has been milk, cheese, bread and vegetables, and this for many centuries. Such a diet, even though it contain no meat, is not a diet low in protein.

On the Texas cattle ranches twenty years ago, although there were thousands of cattle, there was little fresh meat and never any milk. Fresh meat was considered too expensive. Coyotes destroyed the chickens, so eggs were not available. The staple foods were boiled beans, bacon (called sow-belly), soda biscuits, molasses, canned tomatoes, black coffee and a large dish of very greasy flour gravy. The proteins were therefore largely those contained in beans and wheat. The Texan cow puncher raised beef for the inhabitants of the cities but he rarely ate it himself. Yet no one could deny that he was not physically fit. A traveler on arriving at one of the ranch houses called out, "Hello, can I stay here all night?" The answer was usually in the affirmative, and a passing stranger thus received was given the best in the house, always without money and without price, and when he departed was invited to "come again."

The question may be asked about the relative value of beef, veal, mutton and chicken. These are frequently differentiated by the physician in prescribing for his patient. Experiments involving the ingestion of these various forms of meat give no information that one kind is more readily absorbed

than another. Chemical analysis shows no differ-
ence. The substances which are convertible into
uric acid are present in all these meat foods in about
the same quantity. It resolves itself into a question
of habit, personal taste and the local capacity to
cook. In Germany, light colored meats such as
chicken and veal are considered digestible and good
food for the sick. Red meat like beef is deemed
indigestible. A German professor of medicine remarked in his lectures, "Curiously enough, in
America, beef is considered digestible and veal very
indigestible." Veal is well cooked in Germany and
greatly liked. The cooking of beef, on the contrary,
is one of their unsolved problems. These facts,
remembered in connection with the great value of
milk as a protein food, should be held in the mind of
everyone who wishes to inflict his personal dietetic
whims and fancies upon his long-suffering friends.
Ultimately the question is a question of flavor. Man
chooses what he likes best, and, taken in moderation,
what he likes best he digests best. Hence the
demand for meat, which is not necessary and which
could be largely replaced by the cheaper and equally
valuable protein of milk, or even by the proteins of
wheat and of beans taken together.

Rubner points out how the consumption of meat
per head of population in Germany has risen three
and a half fold during a hundred years and now
approximates that consumed in England and in the

United States. In 1813 the consumption of meat in Germany per inhabitant was about at the level of that in Italy today. The large use of meat has come with the concentration of population in the cities. The same transformation has taken place in the United States. In 1820 only 5 per cent of the population lived in cities of 8,000 and over, whereas, in 1910, 33 per cent lived in such cities. This third of the population demands meat in increasing quantity despite its increasing price.

V

THE CURIOUS DISEASE OF BERI-BERI

It has been known for a long time that life cannot
be maintained on an absolutely pure mixture of
salts, fats, carbohydrates and protein. Osborne and
Mendel could not prepare a diet which was chemi-
cally pure in all its constituents and still maintained
life. In most of their successful experiments the
ash of the diet was derived from a powder obtained
by evaporating milk, the protein content of which
had been almost entirely removed. They have
recently reported experiments on rats in which
growth had come to a standstill when the fat in the
diet consisted of lard, but in which rapid growth
ensued when butter fat was substituted for lard.
Butter fat must therefore contain something which
lard does not contain.

The human organism is extremely sensitive to
certain substances in minute quantities. Thus,
epinephrin, the active constituent of the suprarenal
glands, is present in the blood in 1 part in 100,000,-
000 and this is essential to life. One may also
recall the marked influence on growth exerted by
the thymus gland during youth. How profoundly
important the secretions of the internal glands may

become has been vividly demonstrated by Guder-natsch, who gave tadpoles thymus gland and noted their rapid growth and delayed metamorphosis, whereas when thyroid gland was given suppressed growth and premature metamorphosis into pigmy frogs took place.

The profound effect of small amounts of materials formed in the various internal glands is well established. However, it has only recently appeared probable that, in order to maintain the organism in condition, a small quantity of something not hereto-fore recognized must be present in the food.

Eykman, in 1897, called attention to the fact that the disease beri-beri was prevalent among those rice-eating nations which partook of rice prepared in a certain way. The Bengali eats unpolished rice, that is, rice of which the red husk or pericarp has not been removed and he does not suffer from beri-beri. When the native of the Philippines pounds his rice in a large mortar with his own hands, the milled product is never so thoroughly freed of its husk as happens to that milled by machinery and hence these natives are partly protected from the disease. A diet which is based almost exclusively on white rice causes beri-beri with certainty. Eykman also found that when chickens or pigeons were fed on white rice they developed a disease similar to beri-beri. General weakness, with paralysis of the legs and wings, due to a polyneuritis, developed, which could be cured by a change in diet. Here, fortunately, were experimental animals which could be used as a means of finding a cure for the trouble. Unpolished rice, or polished rice when meat or beans were added to the diet, did not induce the disease. Polished rice given with rice polishings prevented the disease.

It has been stated that in Irish and sweet potatoes, and in many common breakfast foods, there is not enough neuritis-preventing substances present to preserve health. Also meat, when sterilized at a high temperature, loses these materials. The desirability of sterilization by boiling of milk for infants has recently been questioned by Funk on account of the liability to destruction of these anti-neuritic substances.

Funk has sought to isolate the substances which prevent beri-beri, which he has termed the vitamines. From 100 kilograms (220 pounds) of dry yeast he prepared 1.6 grams of a crystalline substance which, if given in doses of 4 to 8 milligrams to neuritic pigeons, cures them in from two to three hours. This substance on purification yielded three materials called Substance I, Substance II and nicotinic acid. Substance II was inactive. Substance I and nicotinic acid were severally of little value in curing the polyneuritis of pigeons, but when 3 milligrams of the one and 2 milligrams of the other were administered together, the diseased pigeons were cured in two to four hours. Rice polishings yielded these same two substances.

It therefore appears that a class of vitamines exist in the vegetable kingdom, and that these are necessary to the normal growth and nutrition of animal tissue. They become a part of animal tissue, so that if meat be eaten, their direct ingestion from the plant becomes unnecessary. They enter into normal milk. If the nursing mother has beri-beri the infant acquires it also.

The authorities consider scurvy, the Müller-Barlow disease, and possibly pellagra and rickets as of analogous origin.

Finally, in this connection, it is well to utter a warning to the effect that beri-beri and scurvy do not occur in the United States. They only occur when a one-sided diet deficient in vitamines exists. It is well to remember that the American Association of Tropical Medicine, in May of this year, denounced legislation in the United States against polished rice as unnecessary. Dr. W. P. Chamberlain, Major, Medical Corps, U. S. Army, who worked experimentally on this subject for two years in the Philippines and saw the disease entirely disappear among the Philippine Scouts, advises against any such legislation, and states that the advocacy of unnecessary legislation weakens the authority of the medical profession upon sanitary matters.

It has thus far been shown that nutrition means

fuel for the machinery, new parts with which to repair the machine and minute quantities of vitamines which produce a harmonious interaction between the materials in the food and their host.

VI

CRITERIA OF THE MONETARY VALUE
OF FOODS

The daily press five months ago contained columns of matter regarding the high price of food materials. Not only foods but commodities in general were at the highest prices in thirty years. Some idea of the rise in food prices is indicated below.

	Increased cost in per cent on Feb. 15, 1913, above ave'ge price 1890-99 in 39 cities, U.S.A.	Increased cost in per cent on Feb. 15, 1913, above cost one year before in N. Y. City
Sirloin steak	61	17
Round steak	85	18
Rib roast	63	17
Pork chops	89	24
Smoked ham	69	13
Pure lard	62	10
Hens	67	8
Wheat flour	27	10
Corn meal	56	0
Fresh eggs	63	18*
White potatoes	24	no data
Fresh milk	40	1

Since the efficiency of labor depends upon its energy and constant repair, it is certainly of no

*Decrease.

small moment that the citizen should know how best to maintain the machine at a maximum of efficiency. Not only that, but in times of trouble he should know where to turn to find nourishment in the form which is best and cheapest.

Who will give him this information? Will the manufacturer of canned tomatoes tell him that tomatoes are valueless in his extremity? No, not unless the manufacturer is forced to do so. And how can the manufacturer be forced to give this information? By being compelled by law to label his can: "This can contains x calories of which y per cent are in proteins of grade C." (Consult also Murlin, *Popular Science Monthly*, October, 1913.) If, through the medium of the schools and the press, everyone knew that a man of sedentary occupation required 2,500 calories and a laboring man 3,000 calories and more, no one suffering from want would spend his money for a can of tomatoes which is little else than flavored water.

It has been estimated that a family of five, including the father, a clerk, the mother who does the housework, and three children, nine and six years and one month old respectively, requires 7,750 calories per day. (Lusk: Food at Fifty Cents a Day, *New York Evening Post*, February 8, 1913.) This is probably within 10 per cent of the true value.

To provide a diet containing 7,750 calories, 5 per cent of which were in animal proteins of grade A and 10 per cent of which were in vegetable proteins of grade C (bread), would have cost as follows on January 28, 1913, in the New York markets:

	Cents
Bread + 4/5 pound salt cod	47
Bread + 3/5 pound smoked ham	48
Bread + 4/5 pound cheese	51
Bread + 2 1/2 pounds milk	53
Bread + 1 1/2 pounds loin pork	56
Bread + 1 1/8 pounds leg of mutton	56
Bread + 1 1/4 pounds cod steak (fresh)	58
Bread + 1 1/8 pounds sirloin beef	66
Bread + 1 1/2 pounds turkey	78

If cornmeal, oatmeal, dried beans or rice had been used instead of bread, these prices would have been lower, whereas potatoes would have slightly increased them.

These figures are for the great staples of diet. It is obvious that the cost of fuel for an adult requiring 2,500 calories would be one-third the cost for the family or an amount not exceeding twenty cents a day at the market price of the fresh materials. This being true of the staple products of a dietary, it is obvious that when more than an average of eight cents is expended for 1,000 calories of nutriment, the diet must include luxuries.

The following represents the market price in cents of 1,000 calories in various staple foods:

COST IN CENTS OF 1,000 CALORIES

Glucose	1⅝
Cornmeal	2
Wheat flour	2½
Oatmeal	2⅘
Cane sugar	3⅛
Dried beans	4
Salt pork (fat)	4½
Rice	5
Wheat bread	5⅛
Oleomargarine	7½
Potatoes	7½
Butter	10
Milk	10
Smoked ham	10¾
Cheese	11⅝
Loin pork	12¼
Mutton (leg)	16¼
Salt cod	19½
Sirloin beef	24
Turkey	40
Codfish steak (fresh)	42

This is the cost of nutritive values and when considered in relation to the preceding table, which includes the proper amount of protein for a family, is of the greatest economic importance.

It is evident that if each package of food were sold as containing a stated number of calories, the widely heralded "food value" of Postum, for example, would "fold its tents like the Arabs and as silently steal away."

If the family of five, before mentioned, keeps a servant the amount of food increases from the equivalent of 7,750 calories to 10,250, an increase of 30 per cent. Three servants will double and six servants treble the food bill. One can thus formulate the household requirements:

	Calories	Cost in cents Minimum	Maximum
Poor family	7,750	50	70
Well-to-do	15,500	100	140
Wealthy	23,250	150	210

Whatever is spent above these amounts is paid for waste or for non-essentials in the form of flavors of high price. High cost may also be due to carnivorous indulgence approximating that of the Eskimo.

The variant introduced by indulgence in meat may be thus illustrated:

All the calories are in	Daily cost in cents to furnish Poor family	Well-to-do family	Wealthy family
Wheat bread	41	82	123
Sirloin steak	186	372	558
Eggs	258	516	774
Turkey	312	624	936
	(0 servants)	(3 servants)	(6 servants)

It is not probable that the food values actually consumed are very different in the various well-nourished families. Only the cost can vary enormously.

The cost of eggs in the above table is estimated at the low figure of 24 cents a dozen. At the present moment with the cost of fancy fresh eggs at 75 cents a dozen, it would cost the wealthy family $24 a day if its sustenance were solely derived from eggs. To support the three differently conditioned families on eggs alone would require 129, 258 or 387 eggs daily.

Where there is surplus income and financial limitations do not hamper the expenditure for food, the subject-matter of this paper has no practical bearing. When each individual can partake of nourishment in accordance with the dictates of a normal appetite there is no danger of undernutrition. It is, however, among the poor, the class to which scientific knowledge is the last to reach, that such knowledge would be most valuable. Recently, Miss Dorothy Lindsay made a report concerning the diet of the working classes in Glasgow. Noel Paton, in his preface to the report, asks this important question, "If a suitable diet is obtainable and is obtained, is it procured or can it be procured, at a cost low enough to leave a margin sufficient to cover the other necessary expenses of family life, with something over for those pleasures and amenities without which the very continuance of life is of doubtful value?"

Miss Lindsay examined the dietary habits of sixty

families whose incomes varied from $3.25 to $15.00 weekly. Wherever the wage was above $5.00 a week the family was adequately nourished and the man of the family received 3,000 calories daily, that is to say, enough to make him an efficient machine. Where the wage fell below $5.00 a week there was always undernutrition. The staples of food were bread, potatoes, milk, sugar, beef and vegetables. Little use was made of cheaper oatmeal, peas, beans, cheese or fish. Oatmeal was used in forty-six out of sixty families, but the average amount per man per day was less than one ounce (less than 100 calories). The amount spent for food varied between 62 and 87 per cent of the total wage. In the families showing an average of 87 per cent expenditure for food, the father was a drinker and the family in debt for the rent. It was found that when the weight of a child at a given age was much below the normal, inadequate diet was almost always the cause. Miss Lindsay concludes that one of the main contributing factors of malnutrition among the poor is bad marketing.

The experience of America in the matter of school lunches for the children of the poor, has shown so marked an improvement in their physical and mental well-being, that it is reported to be of demonstrable economy to the state to feed the undernourished children. But the state should also teach the mother the value of bread and milk, and that weak tea cannot take the place of milk in the nourishment of the child.

Recently F. C. Gephart of the Russell Sage Institute of Pathology has undertaken a study into the physiological value of the various portions of food sold over the counter of the Childs restaurants in this and other cities. All the New York restaurants of this company have been visited and of those in the entire country about 70 per cent. This is not the time to give more than a few general statements regarding the results. For obvious reasons, no knowledge of the work has passed beyond the confines of the laboratory circle. Samples have been analyzed for protein, fat and carbohydrate; in some instances also for ash, and the heat of combustion has been determined in each case. Four hundred samples in all have been collected and analyzed. For comparative standards the calories in bread and vintage champagne have been selected, neither of which was purchased in the restaurant. The following table shows a few of the results:

FOOD VALUE OF PORTIONS, IN CHILDS RESTAURANTS,
INCLUDING BREAD AND BUTTER WHEN SERVED

| | Cost in cents | Calories | | Calories for five cents | Cost in cents per 1,000 calories |
		Total	% in protein		
Bread*	5	933	12	933	5
Apple pie.........	5	337	5	337	15
Boston pork and beans	15	828	12	276	18
Ham sandwich...	5	170	20	170	30
Corned beef hash.	15	507	14	170	30
Beef stew........	15	461	25	154	32
Club sandwich....	25	409	20	82	61
Sliced pineapple..	5	36	46	36	138
Tomatoes, lettuce mayonnaise	20	53	16	13	385
Pt. of champagne*	200	345	0	9	588

When one considers this table in the light of the knowledge that the average workman need not expend more than 8 cents per 1,000 calories of energy, it is perfectly evident that Childs restaurant is not a charitable undertaking, but rather an institution for men of moderate means.

The extreme variability of the purchasing power of money for food stands here exposed in the limelight.

It is for this reason that the government should take up this matter. Suppose the pot of Boston

*Not purchased in the restaurant.

baked beans, as sold, were guaranteed to contain
1,000 calories, 12 per cent in protein, the workman
would then know what he was getting for his money,
if he only were taught the simplest elements of the
subject of nutrition.

The government could give information with
regard to all food stuffs sold in packages. The
determination of the heat of combustion of a dried
sample of food takes fifteen minutes. Probably
three hours would suffice to make a complete analysis by a government expert. The manufacturer
should send his sample can to the Bureau of Chemistry at Washington, declaring that to be his standard
and requesting information regarding his label. He
should pay for this analysis as a patentee pays for
his patent. If the government, at any time, should
find the manufacturer selling a material on the
market of different character than the standard
deposited with the government, the manufacturer
should be heavily fined.

Complaint is made by the manufacturers of foods
and patent medicines, that in other countries scientific men sit on their boards of directors and give
advice. In this country this sort of thing is rightly
discountenanced. Food manufacturers have recently requested a group of scientific men to meet
as a board, name their own salaries and give advice.
But the day of the sale of a man's scientific reputa-
tion and that of the institution with which he is

connected has almost passed. The scientist has come to have sufficient altruism to believe that his services belong to all the people and not to a set of money-making individuals.

It is a matter of common knowledge that physiological chemistry is as advanced in this country as anywhere else in the world. There is a great opportunity opened here for good work. Pure food is necessary. Foul ·food should be strangled at its source. But besides this a widespread knowledge of what food really is would be of great value and would blast out of existence some commercial dietary impostures of the meanest description. Appeal is therefore made to the understanding of physicians and of the educated people of this country to take interest in this subject to the end that enlightened activity for the welfare of mankind may follow.

INDEX

INDEX

www.ingramcontent.com/pod-product-compliance
Lightning Source LLC
Chambersburg PA
CBHW070401190526
45169CB00003B/1056